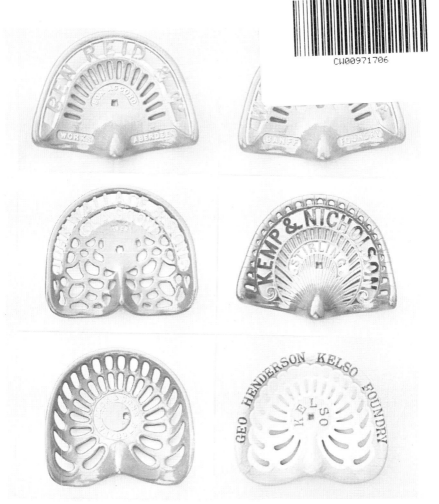

Cast iron seats from Scottish implements made by (from top, left to right): Ben Reid, Aberdeen; Watson Brothers, Banff; John Wallace and Sons, Glasgow; Kemp and Nicholson, Stirling; Jack and Sons, Maybole; George Henderson, Kelso.

SCOTTISH AGRICULTURAL IMPLEMENTS

Bob Powell

Shire Publications Ltd

CONTENTS

Introduction 3
Winter 4
Spring 11
Summer 15
Autumn 21
Transport 29
Further reading 32
Places to visit 32

Copyright © 1988 by Bob Powell. First published 1988. Shire Album 218. ISBN 0 85263 925 2.
All rights reserved. No part of this publication may be reproduced or transmitted in any form or by any means, electronic or mechanical, including photocopy, recording, or any information storage and retrieval system, without permission in writing from the publishers, Shire Publications Ltd, Cromwell House, Church Street, Princes Risborough, Aylesbury, Bucks HP17 9AJ, UK.

Printed in Great Britain by C. I. Thomas & Sons (Haverfordwest) Ltd, Press Buildings, Merlins Bridge, Haverfordwest, Dyfed SA61 1XF.

British Library Cataloguing in Publication Data available.

Cover: *Threshing oats at The Dell, Kirkmichael, Moray, in January 1987, using a portable 'mill' built by R. G. Garvie and Sons of Aberdeen in 1947.*

ACKNOWLEDGEMENTS
I would like to thank Aberdeen City Libraries for the photograph on page 3; Robbie Gordon; Willie Kelly; Moray District Council Department of Libraries, Museums and Archives; Doug Morrison; Ian Morrison, Curator of the Falconer Museum, Forres; Eric Sixsmith; Moira and Charlie Spence at the Ladycroft Farm Museum; Gerald Stewart; and especially my wife, Faye. Special thanks to Dr A. Fenton, Dr J. Shaw, and Ms D. Kidd of the National Museums of Scotland for permission to reproduce photographs from the Scottish Ethnological Archive on pages 5 (lower), 9 (upper right and lower), 10 (lower right), 11, 12 (top), 14 (bottom), 16 (bottom), 20, 21, 22, 23 (top and bottom), 24, 25, 29, 30 (bottom), 31.

Harvesting oats in the Orkneys using a reaper-mower pulled by oxen. The driver uses the rake to clear the cut crop from the cutter bar.

Isle of Skye crofters, about 1880, planting potatoes in 'lazy beds' using the caschrom or foot plough. It was said that twelve men with caschroms could cultivate one acre (0.4 ha) per day. The women are manuring the beds with seaweed, which is carried from the seashore in wicker creels.

INTRODUCTION

This book is about the huge variety of agricultural implements made and developed in Scotland from the early eighteenth century to the early twentieth century. The term implement is used in the broadest sense: there is often no clear distinction between farm implements and farm machines, and an implement is interpreted as being a manufactured item that does a job of work; therefore, a range of items from the caschrom (foot plough) to the tractor is depicted. Only those implements commonly associated with agriculture are included and each category is treated chronologically within a seasonal framework. The old are compared to the improved yet still related to their primary function on the farm.

It is hoped that interested readers will refer to the books suggested under 'Further reading', to gain an historical understanding of Scottish farming systems (for example, run-rig and infield-outfield systems) and the effects of the highland clearances, of increased enclosure, the development of model farms and the formation of agricultural societies. It is also hoped that Scottish readers will be encouraged to record some of the implements made by local manufacturers, particularly while vestigial remains from the horse era may still be found turning up at 'roups' (farm sales).

The 'Old Scots Plough', or 'twal owsen' plough as it was known in Banffshire in the eighteenth century. This rudimentary plough with its flat mould-board did a serviceable job ploughing field 'rigs' (ridges).

WINTER

PLOUGHING

Excepting some highland areas where the caschrom (foot plough) or a light single stilted or handled plough was used, the main Scottish plough, employed from late medieval times until the eighteenth century, was the 'Old Scots Plough'. Often described as cumbersome, the plough was made of wood except for the share, coulter and bridle. The breast or mould-board was wooden and, unlike later improved ploughs, was flat. Consequently, it did not properly turn the furrow but rather broke up the earth. The plough was usually drawn by oxen or by mixed horses and oxen. In Banffshire it was called the *twal owsen* (twelve oxen) plough, though elsewhere it was commonly drawn by eight oxen.

There was no overall successful improvement of the Scottish plough until that of James Small. Born in about 1740, Small was apprenticed in the 1750s to a carpenter and ploughwright at Hutton, Berwickshire. He then became a journeyman in Doncaster, Yorkshire, where he probably had the opportunity to study the improved English 'Rotherham' plough.

Small returned in 1763 to Blackadder Mount, Berwickshire, where he commenced trials to make his 'chain' or 'swing' plough. A clever man, Small claimed to have designed his plough on the best scientific principles and the final product, about 1765, was the result of much experimentation. In his trials he used mould-boards of soft wood in different forms to see how they turned the furrow. Eventually he arrived at a mould-board which curved not only from top to bottom, but also from front to rear. This had the advantage of neatly reversing the furrow slice with the least resistance and friction. Once he was satisfied with the new form of mould-board it was made in wrought iron and later of cast iron from his patterns at the Carron Iron Works, Stirlingshire.

Other improvements were incorporated by Small, including a feathered share that cut the furrow more cleanly, and cast iron sole and land sides. The whole design was lighter, stronger, and could be drawn by two horses. An experiment by the Dalkeith Farming Society showed that in ploughing up old pasture Small's plough needed a force of only 9 to 10 cwt (457 to 501 kg), whilst the Old Scots Plough needed 16 cwt (813 kg). Earlier 'chain' models had a draught chain from the head to a point well back under the beam, supposedly to reduce the strain on the beam, but this was discarded later and the plough became 'Small's Swing Plough'.

Small's plough was a success and by the 1780s was widely adopted in the southern lowlands. At his works Small and his team of about thirty carpenters and blacksmiths produced about 500 ploughs

A twentieth-century single-furrow horse plough made by an unknown manufacturer. Note the large disc coulter and the single stabiliser wheel. Most Scottish ploughs were swing models, having no wheels or only one.

A horse-drawn turnip lifter, or 'ferret' as they are known in parts of north-eastern Scotland. Similar contemporary, late nineteenth-century machines commonly had sledge runners rather than wheels for between the drills.

War. However, it was the onset of the First World War that saw the introduction of the tractor and at this point many larger manufacturers turned to designing tractor ploughs. By 1917 firms such as Sellar of Huntly were producing three-furrow trailed tractor ploughs using the technical knowledge they had gleaned from their previous productions. However, many of the smaller local manufacturers were eventually eliminated because they were unable to withstand the new scales of production required.

LIVESTOCK HUSBANDRY

From the mid eighteenth century advances in animal breeding led to increased livestock numbers and a change in the way they were kept, in particular overwintering. Turnips, beans, whins (gorse) and oilseed cake were all increasingly used as winter feed. This led

Top left: *A blacksmith-made turnip 'hasher'. A turnip placed in the basket of knives was forced through and sliced by pressure placed on the counterbalanced handle.*

Left: *Mechanical turnip cutter made by Kenneth McKenzie of Evanton, Ross-shire. Belt-driven via the fly wheel, the rotating base plate with two cutting blades sliced the turnips, which were held by the raised cast iron sections in the basket.*

Below: *The turnip-cutting cart made by Kemp and Nicholson of Stirling not only transported the roots but also had a wheel-driven geared cutting mechanism for easy field distribution.*

to the development of turnip lifters, turnip cutters or 'neep hashers', hand-mills for grinding, whin bruisers, cake breakers and so on.

Late in the nineteenth century horse-drawn turnip lifters were developed such as that from T. Hunter of Maybole which ran on sled runners and had shares which prised the roots from the ground. The turnips were then collected manually. Fully mechanised harvesting did not occur until well into the twentieth century. There was a variety of turnip cutters, many made by blacksmiths, and their popularity reflected the success of the crop for animal feed.

Above left: *A cake breaker made by George Sellar and Son of Huntly. Cake made from substances such as linseed was a valuable feed supplement for animals. Delivered to the farm in slabs, it had to be broken before being fed to the beasts.*
Above right: *A whin mill at Gartly, Aberdeenshire. Whins (gorse) placed in the stone's channelled path were crushed when the stone was turned by a horse drawing on the end of the beam. They made a valuable, usually supplementary, winter cattle feed.*
Below: *Two-handed broadcasting from sowing hoppers in the Lothian area after the Second World War. Following the sowers are horse-drawn harrows covering in the dispersed seed.*

Left: *Refilling a seed hopper for two-handed broadcast sowing.*

Below left: *The 'Aero' broadcast fiddle sower, made in Kilmarnock, was sold all over the British Isles. It dispensed with the manual skill required by a sower using a seed hopper machines.*

Below right: *The illustration in this advertisement shows a horse-drawn five-coulter seed drill. In some areas, until more recent times, drill sowing was generally disregarded in favour of broadcast sowing.*

BIG CROPS Essential for 1940
Sow Right for Crop Results
THE AERO SOWER

Sows the Seed right, so gets the Crops.

Read these, please:
"A neighbour passed while I was sowing a field to-day, and he says he never saw a better job done. The majority certainly prefer yours."
—*Notts.*

"I may tell you the Broadcaster Seed Sower you supplied is absolutely 'Top Hole' and indispensable on the Farm."—*Norfolk.*

For Grain and Small Seed Sowing

Seed Bag of strong close-woven white cotton canvas

Supplies ready

Cash Price

27/6

Carriage 1/-

No Delay

Outstanding for equal seed distribution. Easy and Speedy Effective Work. One Step per Bow Stroke:
Left-Right: Left-Right.
Built for Wear. Practical, Simple, Efficient. Instantly regulated for all Seeds. Working Instructions affixed to each Sower.
Descriptive Leaflet on request. Thousands sold.
D. LAUDER, 2-6 Portland Street, KILMARNOCK

Telegraphic Address, "SHERRIFF," Westbarns.

The Best 90 years ago . . .
and still the Best to-day . . .

THOMAS SHERRIFF & CO.
AGRICULTURAL ENGINEERS
and MILLWRIGHTS,
WESTBARNS, DUNBAR.
(Established 1818.)

Make the Best Drill and Broadcast Sowing Machines for all kinds of Grain, Seeds, and Manures.

Awarded the Highland Society's Highest Honours at every Competition for the last 60 years, and the U.E.L. Society's latest Award, making a total of 15 Gold Medals, 35 Silver Medals, 8 Bronze Medals, and 54 Money Prizes, in competition with all the leading makers.
May be seen Working on the Best Cultivated Farms over the Civilized World.
MACHINES in their Season by all the Leading Makers Supplied on Best Terms.
Repairs of all kinds of Farm Implements, Thrashing-mills, Engines, and Machinery, Executed by Experienced Workmen.

Sowing grass seed with a broadcast sower, similar to machines made by firms such as Ben Reid and Company of Aberdeen.

SPRING

SOWING

There were two methods of sowing grain (oats, barley, wheat) and seeds (grasses, clover): broadcasting, where the seed was scattered by hand or machine, often on to unbroken furrows before being harrowed in, and drilling, where the seed was directly sown into the ground by machine in regular drills at a regular depth. Both methods were used from the eighteenth century, although broadcasting was favoured in many areas until after the Second World War.

Into the nineteenth century the standard method of sowing was one-handed broadcasting. A doubled linen sheet was partly wrapped around the chest and left shoulder, and supported with the left arm to form a seed holder, leaving the right arm free for sowing.

Two-handed broadcasting was adopted in the nineteenth century. An ovoid or kidney-shaped hopper, formed of linen or hessian on a wooden or metal frame, was strapped to the sower, allowing both hands free for sowing.

The horse-drawn broadcast sower, extensively used in Scotland, was introduced from Yorkshire in 1816 and first made in Scotland by Lowrie of Edington, Berwickshire. In about 1830 Scoular and Sons of Haddington devised an 18 foot (5.5 metre) wide machine that folded into three for road travel and gateway access. Later two- or three-section machines were usually 18 feet (5.5 metres) wide, with the wheel-driven mechanism enabling 3 to 4 acres (1.2 to 1.6 ha) per hour to be sown. By 1850 a three-section machine was made by James Slight of Edinburgh and a jointed machine was exhibited at the 1851 Great Exhibition by James Watt of Biggar. By 1900 many were made by firms such as Ben Reid of Aberdeen, and some tractor conversions are still used today for 'seeds'.

The hand broadcast sowing machine or fiddle sower was introduced in the late 1800s. Supported by the left arm, it had a wooden frame and canvas hopper, from which a regulated seed flow fell on to a segmented wheel. The seed was then spun from the wheel by the manual reciprocating movement of a thonged bow. The sower was popular with smallholders and is occasionally used today.

Credit for inventing the drill sower belongs to the early eighteenth-century English agriculturalist Jethro Tull. By the middle of the eighteenth century Tull's ideas were being experimented with in Scotland, by such as William Craik of

Above: *Nineteenth-century grubber or cultivator made by Jack of Maybole. This model has the tines and shares forged in one piece. Later models, such as made by Sellar and Son of Huntly, had replaceable share points.*

Left: *The double-breasted ridge or drill plough was used to open ridged furrows for planting potatoes or to form ridges on which turnips were sown. The hinged breasts were adjustable to alter the width of the furrow.*

Crofter's partly wooden drill plough on a three-wheeled road trolley. It is smaller and lighter than the standard all-metal drill plough.

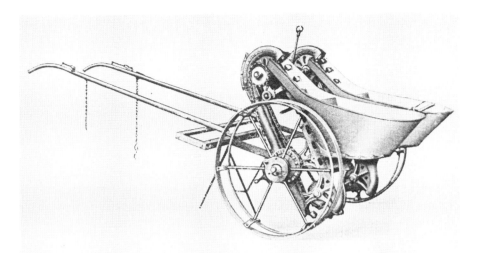

A two-row 'Richmond' potato planter made by Wallace of Glasgow in the early twentieth century. The machine's cage wheels allowed it to run on top of the ridges while the seed potatoes were deposited in the furrows or drills. The potatoes were then covered in by using a drill plough to split the ridges.

Arbigland, Kirkcudbrightshire. Scots drills were produced by the early 1800s. In 1828 S. Morton, an implement maker from Leith Walk, Edinburgh, produced a variable-width three-coulter drill sower. In the 1840s James Slight of Edinburgh produced a lever drill sower based on an English design but at one third of the price, to be attractive to Scottish farmers. Various drills were available by the mid nineteenth century, with five to eleven coulters, at a standard 9 inch (22.85 cm) drill spacing. The six-drill 'East Lothian' design was popular, though it was not until well into the twentieth century that drill sowers were widely adopted.

POTATO PLANTING

After winter ploughing potato land needs cultivating to remove weeds and form a fine tilth for making drills. John Finlayson of Muirkirk, Ayrshire, invented, in about 1820, the first successful Scottish cultivator, with a wheeled triangular frame and several tined shares with variable cultivating depth. Finlayson's initial design was soon improved, though elements were still used a century later.

From the eighteenth to the twentieth century, potatoes were mainly planted manually, in crofter's hand-dug 'lazy beds', and in field drills drawn out and closed again by single-row ridging or drill ploughs. These ploughs were commonly manufactured from the early nineteenth century. Three-row ridgers, made by firms such as T. Hunter of Maybole in the late nineteenth century, were formed on a similar frame to the cultivator but were not as popular as the single plough.

Potato-planting machines were developed in the 1850s. Four Scottish planters were exhibited at Glasgow in 1875, including a two-row planter made by G. W. Murray and Company of Banff which in 1880 was awarded a silver medal. The early machines were not faultless and there was resistance to their use until well into the twentieth century.

Field rollers, whether of wood, stone or metal, crushed clods, consolidated the seed bed and flattened the field for harvesting when using mowers, reapers and binders.
Top: *A late example, about 1900, of a wooden roller made from a tree trunk.*

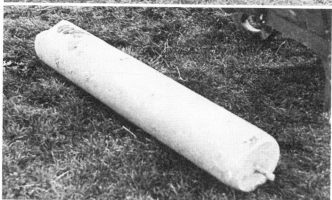

Centre: *A stone roller carved from granite and dating from the late eighteenth or early nineteenth century.*
Below: *Iron rollers, popularised in the nineteenth century, being used at Nether Balfor, Durris, Aberdeenshire, in 1913.*

Driving home a load of hay in about 1910.

SUMMER

HAYMAKING

Haymaking was introduced to Scotland in the early eighteenth century. By the nineteenth century increased livestock numbers, notably of sheep, and often severe weather made haymaking for winter fodder crucial.

Excepting in some crofting areas where the sickle was preferred, grass for hay was mainly scythed until the 1850s, when the mowing machine was developed, primarily by English firms, following the introduction of the reaping machine. In Scotland combined reaper-mower machines such as Jack of Maybole's 'Caledonian Buckeye' were popular. By 1900 mowing machines generally replaced the scythe except on smallholdings and for competition mowing.

After field drying, the mown hay swaths were raked together and forked into small field ricks or haycocks. Horse-drawn rakes were used from the early nineteenth century but most were not Scottish. One popular introduction in 1828, which was first used widely in Scotland, was the American hay rake or 'tumbling tam' as it was later known because of its flip-over action.

The hay was usually taken from the field to the farm steading on hand-loaded hay carts. Hay sledges that could transport a haycock from the field to the steading were developed by the 1880s, though not widely adopted.

At the steading the hay was forked into a stack. Manual forking was easy until the stack was above head height, when a horse-powered fork such as those made by Wallace of Glasgow was useful. This comprised a pole, jib, pulleys and a grab which hoisted the hay above the stack and deposited it. The hay remained in the stack until being fed as winter fodder, or until cut and pressed into bales for sale. Few Scottish hay presses were made.

TURNIP SOWING

Early summer is the usual time for the cultivation and harrowing of the land in preparation for sowing one of Scotland's principal crops, turnips or 'neeps'.

From the late eighteenth century turnips have usually been drill-sown on ridges formed in the cultivated tilth by a drill or ridge plough. This gave a deeper growing soil, and easier thinning, weeding and inter-row cultivation than was possible in earlier times when turnips were commonly broadcast on flat land.

Left: *The scythe with a Y-shaped sned (shaft) was favoured for mowing hay and corn in north-eastern Scotland from the early nineteenth century. The English S-shaped sned was generally favoured in the south, whilst a pole sned was used in some south-western areas.*

Centre: *Hay sledge made by John Wallace and Sons of Glasgow. It was backed up to the field haycock and tipped so that the hay could be winched on.*

Bottom: *Hay bogey or sledge made by A. and W. Pollock of Mauchline, Ayrshire, in the late nineteenth century.*

Above: *A rare, late nineteenth-century, partly wooden, two-drill, horse-drawn turnip sower. Concave front rollers consolidated the planting ridges, while seed from the seed boxes was deposited down spouts to the sowing coulters. The rollers behind firmed in the seed.*

Right: *Hand-pushed, single-coulter sowing barrow of about 1900. Seed in the wheel-driven revolving drum in the seed box was shaken down a spout to a coulter. A roller behind firmed in the seed, often turnip, sown by the farmer replanting ungerminated field drill gaps.*

Soon after the first drill sowing machines were developed in England by Jethro Tull, drilling turnips was attempted in Scotland. Mr Craig of Abbeyland, Dumfriesshire, is acknowledged as being the first to drill turnips, in 1745, whilst Mr Dawson of Harperton, Kelso, is said to have been the first to have sown on ridges, in 1760.

Various drills were made in Scotland using Tull's basic ideas. These ranged from hand-pushed models that sowed one row at a time to horse-drawn two-drill sowers, which became widespread by the end of the nineteenth century. The latter machines were made by firms such as James Gordon of Castle Douglas and

James McKidd's Thurso Foundry Patent Drill Scarifier, developed in the late nineteenth century, was used to weed drills of young turnips. As the young plants passed under the ridge-consolidating rollers, adjustable shares pared weeds from the ridge sides.

An early twentieth-century adjustable wooden drill harrow with five iron tines, used to cultivate between crop rows, such as potato or turnip drills.

Adjustable drill hoe or cultivator for inter-row cultivation (for example, between turnip drills), pulled by one horse. The implement is blacksmith-made.

A hay press made by J. Turnbull and Son, Carnock, Larbert, which won first prize at the Highland and Agricultural Society Show in 1889. Hay cut from the stack was placed inside the press and the levers were then worked to compress it into bales.

Sellar of Huntly. They worked on the principle of seed being fed regularly from a box down a spout to a coulter that split the ridge top. Many of the early twentieth-century machines were converted to tractor hitch and may still be seen in use.

Another sower, dating from about 1730, was the 'Bobbin John', designed by the Udny of Udny, Aberdeenshire. This was a perforated tin can on the end of a stick which was used to shake seed out along the row. The name was applied to several single-row sowers, and especially to a later discoidal design that rolled along the row. It was valued until recently for replanting ungerminated parts of drills that were first machine-sown.

Once germinated, turnips require thinning, normally with a hand hoe. Machines were invented for this purpose, such as that by T. Wardlaw, of Toughmill, Dunfermline, in the late nineteenth century, but have still not become commonplace.

The scarifier invented by James McKidd of Thurso, and copied by others including T. Hunter of Maybole, was used to weed the ridges. Inter-row cultivation was also carried out by the drill harrow or horse hoe, such as those developed by Wilkie of Uddingston in 1820 and others. These implements weeded, softened the soil for hand hoeing, and aided moisture penetration for plant growth.

In 1803 the Highland and Agricultural Society of Scotland offered a prize for a successful reaping machine. This is a model of the first notable attempt, devised in 1805 by Mr Gladstone, a millwright from Castle Douglas, Kirkcudbrightshire.

A model of the first truly successful reaping machine, developed by Patrick Bell from 1826 to 1829. It was driven from behind by two horses.

Reaping oats with scythe-hook sickles at Cramond Brig, Braehead, Midlothian, in about 1920. Attached to the sickle handles are leather straps which loop around the reapers' wrists for additional support.

AUTUMN

HARVESTING

The Scottish harvest from the seventeenth century to the nineteenth was principally reaped by seasonal migrating shearers using sickles.

In the southern lowlands of the eighteenth century the work was often done by women who came from the highlands and islands. Working in a 'bandwin' of seven shearers, they worked along the 'rigs' (field ridges). Six sheared, whilst one bound the sheaves. The crescent-shaped sickle had a toothed edge that sawed through the straw when drawn towards the shearer. Up to 2½ acres (1 ha) per day per bandwin could be cut.

By the early nineteenth century the toothed sickle was slowly replaced by the larger, smooth-edged 'scythe-hook' sickle. It was used principally, by men, with a slashing motion across the rigs and the highland migrants were replaced by teams of Irish labourers who annually arrived in their thousands.

The scythe had long been used for grass mowing but was not popular for harvesting. Despite eighteenth-century attempts to encourage its use, it was believed that the scythe shook too much corn from the ear. However, its use did spread from the early nineteenth century, especially in the north-east. One scythesman could cut 2 acres (0.8 ha) of oats per day, compared with the bandwin's output. The sickle persisted into the nineteenth century, but by the 1850s both sickle and scythe were increasingly replaced by the reaping machine.

The first notable Scottish attempt at a reaping machine was patented in 1806 by Gladstone from Castle Douglas, Kirkcudbrightshire. Drawn by one horse, it had a self-sharpening side rotary cutter. The invention was greeted with enthusiasm but the complex design was a failure.

In response to a £500 prize offered by the Dalkeith Farmers Club in 1812, the second notable design was submitted by James Smith of Deanston. Smith's 20 foot (6 metre) long machine had a rotary blade which cut a 4 foot (1.2 metre) wide

Nineteenth-century manual delivery reaper made by Jack of Maybole, Ayrshire. The operator had to drive the horses and rake the crop back away from the cutter.

swath and was pushed by two horses. An improved model submitted in 1813 to the Dalkeith Farmers Club impressed them enough to present him with a piece of silver plate worth 50 guineas (£52.50). Favourable reports by the Highland and Agricultural Society gained him a second similar award. Smith improved his machine until 1835, when it succeeded in trials at the Highland and Agricultural Society's Ayr Show. In spite of this achievement Smith's machine was developed no further. It was considered too long, heavy, unwieldy and costly.

The breakthrough came in 1826 when Patrick Bell, later Presbyterian minister of Carmyllie, Angus, made a model reaper with a fingered cutter bar based on the clipping motion of garden shears. Bell presented his model to Sir John Graham Dalyell, a mechanic, who persuaded him to make it full size.

Afraid of testing his machine openly, Bell first tried it in a barn in 1828, where he successfully cut an oat crop that he had secretly planted. Even so, his first outside trial was at night. Bell then tested his machine publicly. Following one trial at Gladstone Farm, Monikie parish, in 1829, a copy of his machine was made at the East Foundry, Dundee, and presented to the Highland and Agricultural Society, who rewarded Bell with a £50 premium.

Bell's reaper was pushed from behind by two horses. Gearing from the wheels drove the 5 foot (1.5 metre) cutter bar of thirteen oscillating upper and twelve fixed lower shears. Additional gearing powered the revolving reel which drew the corn to the shears and also aided the cut corn to fall on the endless belt. This belt deposited the corn in a swath to either side of the reaper by the moving of a clutch. The reaper was capable of cutting 12 acres (4.9 ha) a day.

There was resistance to mechanical reaping and by 1832 only ten of Bell's reapers were working in Scotland. Four reapers were sent to America, where it was said Cyrus McCormick based his successful machine on Bell's principles. Bell, like James Small, never patented his ideas, wanting all users to benefit freely.

Threshing with flails in the Orkney Islands in the late nineteenth century. The sheaves are placed on a ground sheet to collect the grain as it is beaten out.

THRESHING

Early Scottish threshing, the act of separating grain from the cut corn, was carried out using the flail. The flail comprised a 5 foot (1.5 metre) handle or handstaff, a simple universal joint often made of leather and a 3 foot (0.9 metre) beater or 'souple' made of wood, rope or even rolled sealskin. The beater had to be swung skilfully and brought down on the sheaves laid on the threshing floor. Despite the acceptance of the threshing machine in the early nineteenth century, the flail continued in use, especially in the highlands and islands, into the twentieth century.

Credit for the first attempt to mechanise threshing is usually given to Michael Menzies from East Lothian, who, in 1732, devised a water-driven machine. Comprising flails fixed to a revolving beam, it was not a success as it broke easily.

The next notable machine was made in 1758 by Mr Stirling from Dunblane, Perthshire. Corn was fed into the top of a cylinder, measuring 3 feet 6 inches by 8 feet in diameter (1 by 2.4 metres), that enclosed a water-powered vertical shaft with four beaters. It successfully threshed oats, but not wheat and barley.

One of the greatest Scottish contributions to agricultural improvement came when Andrew Meikle, from Tyninghame, East Lothian, started experimenting with threshing machines at Knowmill. Meikle based his ideas on those of Mr Ilderton of Northumberland, and in 1786 he designed his successful threshing machine or 'mill', in which the corn was fed between two fluted rollers on to a revolving drum with beaters. Soon after its invention a fanner was added to separate the chaff from the corn. Meikle's thresher cost about £80, could be worked by water or horse power, and threshed about 20 to 40 bushels (160 to 320 gallons; 727 to 1455 litres) per hour. It represented a good saving for farmers over manual threshing and was widely adopted in Scotland by the early nineteenth century. One maker, Gladstone of Castle Douglas, had erected two hundred 'mills' on Meikle's principles by 1810. Meikle failed to enforce his 1788 English patent for his machine and subsequently a public subscription was raised to provide for his old age. He died in 1811, aged 92.

By the mid nineteenth century threshing mills were adapted to steam power and at the end of the century to oil

A hand- and foot-powered threshing machine advertised by Garvie of Aberdeen in 1920. These machines were advertised as being suitable for smallholdings, where the crofter often had no mechanical power.

engine power. Both stationary and portable mills were made in all sizes by 'millwrights' all over Scotland. Several firms were based in the north-east. Robert G. Garvie and Sons of Aberdeen claimed to have the largest thresher works in Scotland and produced models ranging from hand or pedal mills to large 4 foot 6 inch (1.4 metre) wide drum mills. Similar mills were also built by Barclay, Ross and Hutchison of Aberdeen and McRobert of Turriff. Smaller mills were the forte of firms such as Murray of Banff, Shearer of Turriff, Watson of Banff, Godsman of Aberlour and Ben Reid of Aberdeen.

Stationary mills were favoured in Scotland and most farms had an integral mill, suited in size to their needs, in a barn. Portable mills were usually the choice of a contractor.

WINNOWING MACHINES

After threshing the grain needs to be cleaned or 'dressed' by winnowing or fanning to separate it from unwanted chaff and dust. Traditionally winnowing was often carried out by allowing the grain to fall from the worker's hands in a draught or cross wind, which separated the lighter unwanted dross.

The advance came in 1710 when a wheelwright, James Meikle, father of Andrew, was sent to Holland from Saltoun by Andrew Fletcher to find out about pearl barley mills. Whilst there Meikle observed and subsequently introduced to Scotland the fanner or winnowing machine. Four rotating vanes created a wind which separated the grain from the rubbish as it passed through the machine. It was not immediately popular and people declared Meikle was making 'the Devil's wind'. The fanner seems to have died out for it was not resurrected until 1733, when Andrew Rogers, a farmer from Cavers near Hawick, found one discarded at Leith. By 1737 Rogers made another and by 1740 he was supplying them at £3 each.

The early fanners required the winnowed grain to be manually sieved to remove additional unwanted or small seeds; integral interchangeable sieves were added later. The fanner soon became an incorporated part of threshing machinery, but separate fanners also continued to be built to dress grain for sale in the markets.

POTATO HARVESTING

In nineteenth-century Scotland potatoes were frequently raised by teams of people, often women, using graips or potato forks and working along the drills. Another method was to use a plough, sometimes the drill plough used for planting, but often a raising plough. The raising plough had fingers that replaced the mould-boards so that the earth fell

An Orkney sledge cart drawn by two oxen. Not having shafts, it needed four w , to keep it horizontal. The solid wheels were said to have been cut from ship's hatches.

TRANSPORT

Not until the 1751 Turnpike Road Act did the improved cart become popular in Scotland. Previously tracks and rough terrain made efficient wheeled transport impractical. Loads, varying from dung to fire peats, were frequently carried on the backs of people or ponies in baskets or creels made of woven available natural materials such as rushes, birch or heather. When carried by people they usually had a strap around the shoulders; on ponies, paired creels were hooked on to a simple wooden pack-saddle.

Where the land was less unkind, simple, often home-made vehicles were used. In 1754, an English traveller, Edward Burt, described some in his *Letters from a Gentleman in the North of Scotland*. The wheel-less slipe or sledge was formed of two elongated shafts whose furthest ends dragged on the ground like a North American Indian *travois*. The slipe could either have a platform or a creel attached for load carrying. Another (probably early eighteenth-century) adaptation was the addition of solid wooden wheels to the ends of the slipe shafts to form the *kellach* cart. A similar vehicle was the rung cart which had ladder sides to retain bulky loads.

A further vehicle of the pre-improvement period, favoured in the lowlands, was the tumbler. This was a form of low box cart, with two solid wooden wheels that revolved with, rather than on, the axle. The tumbler was shafted and drawn by horses. This contrasts with the wain, a less common vehicle, which was frequently drawn by paired oxen via a single draught pole.

The eighteenth-century road acts allowed the adoption of improved carts with spoked and dished wheels that turned on the axle. These were already in use in England, but the English four-wheeled wagon was never successful in Scotland because of the hilly terrain. Writers and improvers such as James Small, inventor of the swing plough, encouraged others when he wrote his 1784 treatise on 'Wheeled Carriages' and produced carts at his Leith workshops.

An early twentieth-century box cart which was originally used at Aberlour, Moray. Made principally from larch wood, these carts were capable of being dismantled by unbolting all their main components.

A Midlothian harvest or long cart. The cart sides are extended by frames to increase the load capacity, while open metal wheel arches kept the load, sheaves or hay, off the wheels.